Living and Nonliving Things

Printed in México

ISBN-13: 978-0-15-362008-9
ISBN-10: 0-15-362008-0

2 3 4 5 6 7 8 9 10 805 16 15 14 13 12 11 10 09 08

Living Things

living frog on a frog statue

Living things need food, water, and air.
Living things grow and change.
They can make new living things.

Animals are living things.
Plants are living things.
Living things are almost everywhere on Earth.

Nonliving Things

Nonliving things do not need food.
They do not need water and air.
They do not make things like themselves.

Food and Water for Animals

Animals need food to survive, or stay alive. Animals also need water to survive.

Animals Need Air

All animals need oxygen.
Fish use gills to get oxygen from water.
Some other animals use lungs.

Space and Shelter

prairie dogs

Animals need space to find food.
They need space to care for their young.
Animals also need shelter.
Shelter is a safe place to live.
Prairie dogs dig holes for shelter.

Water, Light, and Air for Plants

Plants need water, light, and air to live. They also need nutrients from the soil. Plants use these things to make food.

Some kinds of plants need a lot of water. Other kinds, like cactus plants, need less water.

sunflowers

Some plants need more light than others.
Sunflowers need a lot of light.
Ferns do not need a lot of light.

Room to Grow

As plants grow, they need more space.
The roots get bigger and longer.
The stems and leaves also grow.

Vocabulary